Bring Science Alive! Grade 5

Unit 2
Earth Systems
Science Journal

Unit 2

Earth Systems

California's aqueducts have been drying up. Rain has been scarce for years. What is causing this drought? As an intern for the EPA, investigate California's droughts to learn how Earth's systems interact with one another.

 Engineering

Anchoring Phenomenon

Think about this unit's **Anchoring Phenomenon**: *For over five years, California's aqueducts had been drying up, and rain had been scarce. This led to a drought.* Complete the chart.

- List what you **know** about this unit's phenomenon.
- Write questions you **wonder** about this phenomenon.

Know	Wonder

Unit Checkpoints

As you complete each lesson, look for this icon ☑ and return to record what you've learned in the lesson.

Lesson	What I Learned
1 What Are Earth's Four Systems?	
2 How Do Earth's Systems Produce Weather and Climate?	
3 How Do Earth's Systems Change Earth's Surface?	
4 How Do Farming and Industry Affect Earth's Systems?	
5 How Do People's Everyday Lives Affect Earth's Systems?	
6 What Can People Do To Protect Earth's Systems?	

Using what you learned in this unit, explain the unit's **Anchoring Phenomenon**: *For over five years, California's aqueducts had been drying up, and rain had been scarce. This led to a drought.*

Claim	
Evidence	
Reasoning	

Lesson 1

What Are Earth's Four Systems?

©Teachers' Curriculum Institute

Observing Phenomena

Discuss: How does a lake look different in the summer versus in the winter?

Observe this phenomenon: *Lakes drain during years of low rain.*

Find It!

Find a nearby lake, river, or other natural water feature. Do you think it drains during times of low rain, too? Take pictures during summer and winter. What differences do you observe?

Think of what you already know about Earth's four systems. Write questions you have.

Modeling Earth's Systems

You will build a model of Earth that includes Earth's four systems. Your teacher will provide you with the materials you need.

Draw a diagram of your Earth model. Label Earth's four systems. Then, find three places on your diagram where two different systems meet. Explain what might happen when these systems interact.

Think of something on Earth that you are not sure which system it is a part of. Which systems do you think it might be part of? Why?

Vocabulary

Match the term to the correct definition.

Word Bank

hydrosphere   biosphere   atmosphere   geosphere

_____ 1. The Earth system that is made up of a thin surface layer of rock, soil, and sediments as well as the materials that are inside Earth.

_____ 2. The Earth system that includes all of Earth's water.

_____ 3. The Earth system that is made up of a mixture of gases that is air.

_____ 4. The Earth system that includes all the living things found on Earth.

My Science Concepts

Reflect on your understanding. Draw an X along each line.

Earth's four systems are the geosphere, hydrosphere, biosphere, and atmosphere. Creating a model can help you understand where you can find different parts of these systems on Earth. The model introduces you to how these systems work together.

still learning   **know it**

Graphing the different sources of water can help you compare the amounts of types of water there are. For example, Out of about 1,400 billion cubic kilometers of water found on Earth, only 13 billion cubic kilometers are easily available to drink.

still learning   **know it**

1. Earth Has Four Systems

You may have planted bean seeds in a garden. When seeds sprout, roots grow into the soil to search for water that the new plant needs. Leaves form on the plant that take in carbon dioxide gas from the air and use it to make food.

You have just read about parts of Earth's four systems: the *atmosphere, hydrosphere, biosphere, and geosphere*. Above Earth's surface is the **atmosphere**, a mixture of gases that surrounds Earth. This mixture makes up the air that you breathe and that plants use to make food.

Earth's water makes up the **hydrosphere**. The hydrosphere includes all the liquid water in oceans, rivers, and lakes. It also includes the liquid water found in the ground. Earth's frozen water in glaciers and polar ice caps are part of the hydrosphere. There is even water in the atmosphere in the form of clouds and a gas.

The **biosphere** includes all the living things found on Earth. You are part of the biosphere. So are the plants in your garden, the bacteria that live on your skin, and the earthworms that live in the soil.

Earth's **geosphere** is made up of a thin surface layer of rock, soil, and sediments as well as the layers of hot molten, or liquid, rock that are inside Earth. Many objects you use every day, such as plastic or metal, come from the geosphere.

The four systems interact with each other to affect Earth. Plant roots growing into soil is an interaction between the biosphere and geosphere. Soil soaking up rainwater is an interaction between the geosphere and hydrosphere. You will learn more about each of Earth's systems in this lesson.

A plant is part of the Earth's biosphere. The soil is part of the geosphere, and the rainwater it absorbs with its roots is part of the hydrosphere. The plant is also using air from the atmosphere.

Label the four different Earth systems in this image.

Write a brief explanation of what you are labeling and why whatever you have selected is part of the system. Include these terms: **atmosphere**, **biosphere**, **hydrosphere**, and **geosphere**.

2. The Atmosphere

Suppose you are on a shuttle to outer space. You pass through Earth's atmosphere made of air. As you move away from the surface, the air's pressure and temperature change.

Air pressure is how much the air pushes on any surface. Air is made of tiny invisible pieces of matter. The closer together the pieces are, the higher the air pressure. The pieces are closest together near Earth's surface because Earth's gravity pulls air down. As you move away from Earth, the pieces of matter in the atmosphere are more spread out. The amount of air decreases as you move up into the atmosphere.

Temperature decreases as you move farther from Earth's surface. But air does not stay cool. It changes in layers.

Layers of the Atmosphere

As you move upward through the atmosphere, the temperature changes. But it does not change in a constant pattern. Instead, it cools, warms, cools again, warms again, and then cools again as you go higher. Scientists divide the atmosphere into different layers depending on how the temperature changes.

The layer closest to Earth's surface is called the troposphere. You live in the warmest part of the troposphere. You live in the warmest part of the troposphere because the sun warms the Earth's surface, which then warms the air above it. As you move up in the troposphere, it grows colder.

Eventually you reach a point where the temperature begins to increase as you get farther from Earth's surface. This is a new layer called the stratosphere. When the sunlight hits the air at the top of the stratosphere, it forms a new substance called ozone that absorbs energy from the sunlight and heats up the air at the top of this layer.

Earth's atmosphere is the mixture of gases that surrounds Earth. It has five layers that vary in thickness and in temperature.

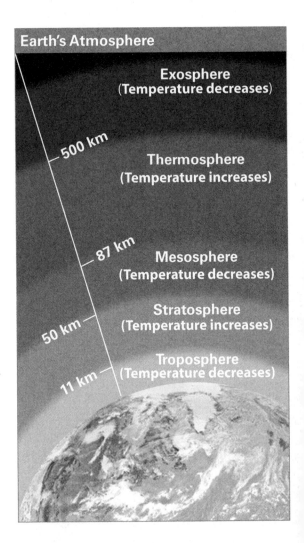

Earth's Atmosphere

Exosphere
(Temperature decreases)

500 km

Thermosphere
(Temperature increases)

87 km

Mesosphere
(Temperature decreases)

Stratosphere
(Temperature increases)

50 km

Troposphere
(Temperature decreases)

11 km

There are several more layers above the stratosphere. There is another layer where the temperature decreases as you move away from the Earth. Above that is a layer where the temperature increases. The temperature decreases and increases in different layers until you reach outer space.

Gases in the Atmosphere

Air is a mixture of gases. The amount of each gas changes slightly from place to place. Nitrogen and oxygen make up almost all of the air in the lower layers of the atmosphere. The rest is made up of other gases such as argon, carbon dioxide, hydrogen, and ozone.

Many of the gases in the atmosphere are important to organisms in the biosphere. Plants use carbon dioxide as a building block to make food. Animals use oxygen to release energy from their food. Carbon dioxide also absorbs energy and warms the atmosphere.

Earth's atmosphere contains many different gases. Some of these gases, such as carbon dioxide, help keep Earth warm so life can exist on Earth.

The Atmosphere Heats Earth

Space is too cold for organisms to live in, but Earth is warm. How does life exist on Earth? Energy carried by light is absorbed by the Earth's surface. Earth then emits, or sends, energy outward. Most of that energy gets absorbed by gases in the atmosphere such as carbon dioxide. The atmosphere then emits some of that energy out into space. It also emits some of it toward the ground, making Earth's surface even warmer. This is called the greenhouse effect. Life can exist on Earth because the greenhouse effect warms the surface. But too much carbon dioxide in the atmosphere causes global warming, the overheating of the atmosphere.

The atmosphere contains the air you breathe and helps to keep you warm. What would happen if Earth did not have an atmosphere?

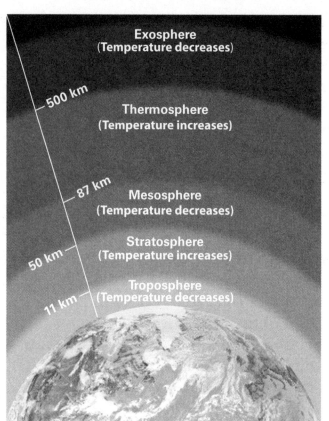

Exosphere
(Temperature decreases)

500 km

Thermosphere
(Temperature increases)

87 km

Mesosphere
(Temperature decreases)

Stratosphere
(Temperature increases)

50 km

Troposphere
(Temperature decreases)

11 km

3. The Biosphere

If you go to a park, you see many living things—trees, dogs, insects, grass, people. If you went under the sea, you would see other living things—fish, sponges, seaweed, maybe even a whale.

This beaver is cutting the tree with its teeth.

Parts of the Biosphere

The biosphere is made up of all the living things on Earth. Living things can be found in a layer that extends from the ocean floor upward several kilometers into the atmosphere. This may sound like a large space with a huge number of organisms. However, the biosphere has the least amount of matter of Earth's four systems.

Animals Affect Earth's Systems

Although the biosphere is small, it has a huge impact on Earth. Parts of the biosphere interact with other systems to change the Earth. Even a single animal can change how the Earth looks. A beaver is an animal that lives in and near streams and rivers. Beavers cut down trees to build a dam made of branches and mud. A dam blocks the flow of water in the river and creates a pond or small lake upstream. Beavers build their home in the center of this pond.

A beaver dam changes a river or stream into a pond. Below the dam, little water flows.

How does a beaver dam affect the environment? Cutting down trees causes soil to be washed away because tree roots no longer hold it in place, changing the geosphere. When the dam is built, water upstream floods fields and forests. The hydrosphere is changed as the river becomes a pond. Many animals lose their homes, but different animals move into the changed environment. The biosphere has also been changed.

Humans Affect Earth's Systems

Humans are part of the biosphere. Like beavers, humans build dams. These dams are much larger and have a much greater impact on the hydrosphere and geosphere. The large body of water that forms above the dam affects the atmosphere. Water evaporates and then condenses, forming clouds. If the body of water is large enough, it may change weather patterns and even the *climate* of the area.

Human activities affect the atmosphere and the hydrosphere. Cars, trucks, and buses send carbon dioxide and harmful chemicals into the atmosphere. You have read that carbon dioxide absorbs energy and raises the temperature of the air. Some of the chemicals that enter the atmosphere combine with water droplets to form acid rain. When acid rain falls to Earth, it can eat away rock, part of the geosphere. After long periods of time, the shape of the land changes. Other chemicals produce ozone, a gas in the atmosphere. Ozone is naturally found in the stratosphere, where it protects living things against harmful radiation from the sun. But ozone in the troposphere is harmful. It causes smog and makes breathing difficult.

This huge open-pit copper mine in Arizona has changed the shape of landforms in the area. Much of the exposed soil will be washed away by rain.

Human activities affect the geosphere. When millions of trees are cut down or minerals are mined from the earth, the soil loosens and is easily washed away by rain. Removing oil and natural gas from Earth can even cause earthquakes, which can change Earth's surface within seconds.

List the different Earth systems you can see in this image. Be specific, and give examples to support your claim.

Describe the impact that the beaver that built this dam has on at least two other Earth systems. Include two of these terms: **atmosphere**, **biosphere**, **hydrosphere**, and **geosphere**.

4. The Hydrosphere

When humans first looked at Earth from space, they called the planet the "blue marble." It really looks like a marble with swirling white clouds, and it definitely is blue.

What makes Earth appear blue from space is its water. Earth is the only planet in its solar system that has water on its surface. About three-fourths of Earth's surface is covered with water. Nearly all of the Earth's water, about 97 out of 100 liters is salt water in the oceans. The rest, less than 3 out of 100 liters is fresh water.

When astronauts first saw Earth from space, they thought it looked like a big blue marble.

Earth's Fresh Water

Many living things in the biosphere need fresh water to survive. But about 69 out of 100 liters of Earth's fresh water is frozen in ice caps at the North and South poles and in *glaciers*, large bodies of slowly moving ice. Living things do not use this frozen water because it takes too much energy to melt the ice.

Surface water and groundwater provide most of the fresh water that living things need. About 30 out of 100 liters of fresh water is underground. Only about 1 out of 100 liters of Earth's fresh water is surface water in rivers, streams, lakes, and ponds.

Streams are narrow bodies of flowing water. They collect water from rain or melting snow or ice. Streams usually flow into larger bodies of water such as rivers and lakes. Rivers are usually wider. Water in rivers eventually flows into seas or oceans. The water in lakes and ponds may have currents caused by wind, but it does not flow.

About three-fourths of Earth's surface is covered with water. Almost all of this water is found in oceans. Most of Earth's fresh water is frozen.

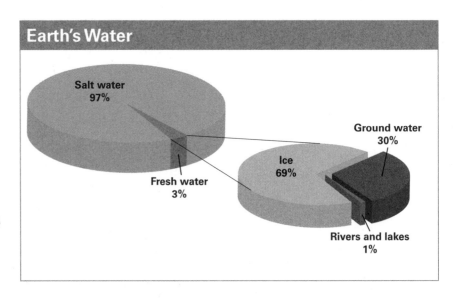

Earth's Water

Salt water
97%

Fresh water
3%

Ice
69%

Ground water
30%

Rivers and lakes
1%

Earth's Salt Water

The ocean holds almost all of Earth's available water. Although different parts of the ocean have different names, they are all connected to form one huge body of water. Ocean water is salty. Most of the salt in ocean water is the same kind of salt you sprinkle on your food. Many other kinds of salt are found in ocean water, too. A measure of the amount of salts in ocean water is called *salinity*.

The temperature and salinity of ocean water vary from place to place. The water is warmest at the surface. At the ocean floor, the water is very cold in most places. Some parts of the oceans are much saltier than other parts.

Oceans are home to thousands of different organisms in the biosphere. These marine organisms interact with the ocean. Some animals swim from place to place to catch food. Fish, whales, and turtles are some of the many animals that swim. Other animals such as sea stars crawl on the ocean floor. Sponges and clams stay in one place and filter tiny food organisms out of the water. Organisms that use photosynthesis to make their own food, such as seaweeds, can live only near the surface because they need sunlight.

Earth's oceans hold almost all of Earth's water. Even though the temperature and salinity of the water vary from place to place, they are all connected and cover most of Earth's surface.

The ocean interacts with the geosphere. Ocean waves beat against land, changing its shape and washing it away. You will learn in the next lesson about how water in the hydrosphere interacts with the atmosphere, affecting weather and climate.

Read the information in the box, and follow the directions.

- 97/100 of Earth's water is salt water in the ocean.
- The other 3/100 is fresh water.
- Most of the fresh water on Earth, 69/100, is frozen in glaciers.
- Of Earth's fresh water, 30/100 is underground.
- The remaining 1/100 of Earth's fresh water flows freely on the surface.

Look at the graph below. Write in the number of liters out of 100 for each bar. Use these numbers:

- 97
- 2.07
- 0.90
- 0.03

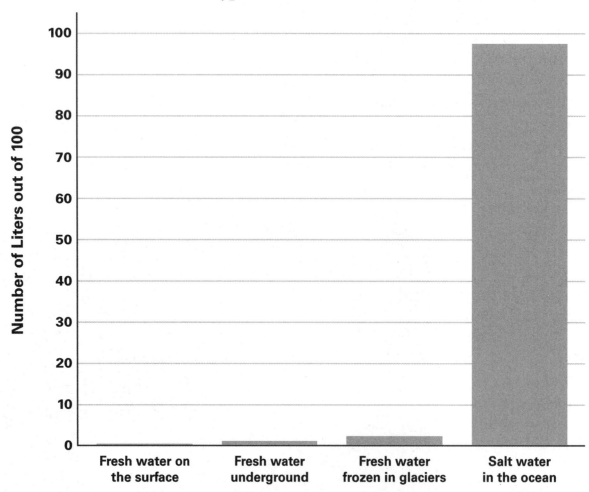

Types of Water on Earth

5. Earth's Water Cycle

Water on Earth does not remain fresh water or salt water. Water is always moving. Sometimes you can easily see the movement of water, such as when a river flows into a sea or when rain falls from the atmosphere. Other times it is harder to see the movement of water. Water is always moving on land and even in the air above.

Water Is Conserved

Water exists in three states: liquid, solid, and gas. The liquid state is usually just called water and solid water is called ice. The gas state of water is known as **water vapor**. Water can change from one state to another as it warms and cools. Putting water in the freezer changes liquid water to ice as the water freezes. A puddle of liquid water will evaporate, or turn into water vapor. Water vapor condenses in the atmosphere to form clouds made of droplets of liquid water.

All water is part of Earth's hydrosphere, and water is never lost from the hydrosphere. Rather, water changes state over and over again as it interacts with the geosphere and the atmosphere. This continuous movement of water over the land and above in the air is called the **water cycle**.

Water is continuously flowing. Sometimes it is easy to see the movement of water, such as when this river flows into the ocean.

The Water Cycle

Water is continuously moving and changing states in a cycle, even when it appears to be still. It evaporates and then falls back to Earth's surface over and over. The energy in sunlight evaporates the water from oceans, rivers, and lakes, turning the liquid water to water vapor. Water vapor is also given off by plant leaves. Water evaporates as water vapor, which rises into the atmosphere.

As water vapor rises, the air cools and water vapor begins to condense into tiny water droplets. These water droplets form clouds. Within a cloud, the water droplets join and grow larger. When the droplets are large and heavy, they fall to Earth's surface. This is **precipitation**, water that falls to Earth's surface. Precipitation has many forms. Precipitation falls as rain where the air is warm and falls as snow where the air is cold.

Some precipitation soaks into the soil, and plant roots take up small amounts of the water. Some runs along the surface as runoff. Some precipitation, when the air is cold, remains frozen on the mountaintops and streets as ice and snow until it melts and joins the runoff. The runoff flows as surface water into freshwater streams and rivers. Eventually, these rivers flow into lakes and oceans.

When rivers run into the ocean, the fresh water mixes with the salt water. The *brackish* water that forms has more salt than fresh water, but less salt than ocean water. As the water continues to run towards the ocean, the water grows more salty until it becomes salt water.

The water in the soil, streams and rivers, lakes, and oceans evaporates. Water vapor is given off by plant leaves. The water vapor rises to the atmosphere again. The cycle continues and repeats again and again.

Water cycles from land to air and back again, but it is never lost from the hydrosphere. The continuous movement of water over the land and above in the air is called the water cycle.

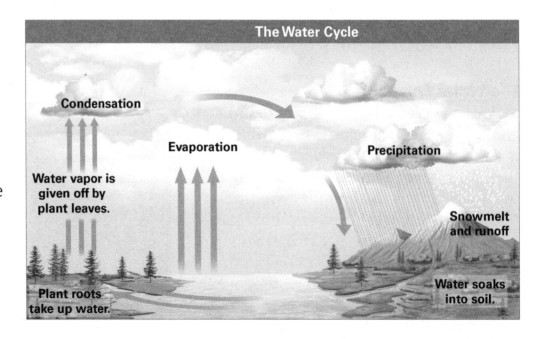

The Water Cycle

Condensation

Evaporation

Precipitation

Water vapor is given off by plant leaves.

Snowmelt and runoff

Plant roots take up water.

Water soaks into soil.

Using the image below, draw and label the water cycle. Include these terms: **evaporation**, **condensation**, **precipitation**, **runoff**, and **water vapor**.

Using the water cycle, explain why water on Earth is conserved.

6. The Geosphere

You may eat a juicy peach. The skin of the peach is its thin outer surface, inside the skin is thick juicy flesh, and at the center of the peach is a hard seed made of two layers. You can think of Earth's geosphere as the parts of a peach.

Earth's Layers

The geosphere includes Earth's surface and its interior. It includes rocks and minerals, soils and sediments, and the processes that shape the surface. The geosphere has four layers: the crust, mantle, outer core, inner core.

The outside "skin" of Earth is the top layer of the geosphere, called the crust. You live on the crust. It varies in thickness from place to place but is generally from 8 to 40 km (5 to 25 mi) thick. The crust is made of solid rock and, near the surface, soil. As you move downward from the surface, the crust gets warmer.

Below the crust is the mantle, the second layer. The mantle is about 2,900 km (1,800 mi) thick. Most of it has a temperature of about 3,000°C (5,400°F) and is made of rock. But there are small parts near the surface made of melted, or molten rock. The mantle is like the thick juicy flesh of the peach.

Finally, beneath the mantle is Earth's metal core. You may have noticed a peach pit has two layers, a brown outer layer and a pale white inner layer. Just like a peach pit has two parts, Earth's metal core has two parts, an outer core and an inner core. The outer core is made of liquid metal. It is about 2,300 km (1,400 mi) thick and has a temperature of about 5,000°C (9,000°F). The inner core is a solid metal sphere with a radius of about 1,200 km (750 mi). It has a temperature of more than 6,000°C (11,000°F).

Earth's geosphere has four layers. Each layer has a different temperature and thickness and is made of different materials.

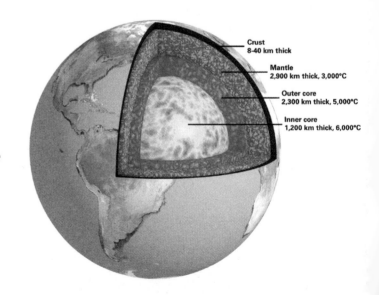

Crust
8-40 km thick

Mantle
2,900 km thick, 3,000°C

Outer core
2,300 km thick, 5,000°C

Inner core
1,200 km thick, 6,000°C

Rocks and Minerals

If you were to examine a rock with a magnifying glass, you would see that it is made up of tiny grains. Each grain is made up of a mineral. Rocks are mixtures of minerals. You are probably familiar with many of them. Table salt, the graphite in your pencils, chalk, metals, and diamond are minerals. Humans use many minerals from the geosphere to make things they need.

Many kinds of rock make up the geosphere. There are three main kinds of rock that form in different ways. *Igneous* rock forms when melted rock from Earth's mantle comes to the surface in a volcanic eruption. On the surface, the rock cools and hardens.

Sedimentary rock forms when tiny pieces of sand, rock, and shells settle and are buried and pressed together in layers. These tiny pieces are called **sediments**. Sediments form when larger rocks and shells are worn away. The sediments are carried by flowing water in a river or stream. When the water slows down, the sediments drop to the bottom. Over millions of years, the layers sediments form sedimentary rock. You can see these layers because each layer contains different minerals.

Metamorphic rock forms from existing rock. Heat and pressure inside Earth's crust change igneous or sedimentary rock into metamorphic rock by squeezing and heating it. Thus, one kind of rock can be changed into another kind.

Rock is a part of the geosphere. Sedimentary rock often has colored layers because the rock forms from layers of sediments that are pressed together.

List the different Earth systems you can see in this image. Be specific, and give examples to support your claim.

This image shows sedimentary rock. Sedimentary rock is formed by sediments pressed together over millions of years. What are sediments, and how can you tell that this rock is made from them?

7. Soil

Soil is the loose material that covers Earth's surface. It contains parts from all four of Earth's systems, such as tiny pieces of rock, minerals, and decayed plant and animal matter. Soil also contains water and air. The bits of rock and minerals came from larger rocks that were worn away. Decayed plant and animal material, called *humus,* came from organisms that died and were eventually covered by more material. It takes thousands of years for soil to form from these materials.

Different parts of Earth's surface have different kinds of soil. For example, forest soil has a lot of humus because many plants grew and died in the forest. Desert soil has a lot of minerals but little humus because few organisms lived in desert areas. Soil on mountains is very rocky, because rainwater washes away much of the humus from mountain soil.

Soil is made of layers. A typical forest soil has an upper layer of humus and plant and animal material that is still decaying. Below this layer is topsoil. Topsoil is loose and contains a lot of humus and minerals. Topsoil holds water well. Most of the soil's organisms, such as fungi, earthworms, and insects, live in topsoil. Seeds and plant roots grow in topsoil.

Under the topsoil is a layer of subsoil, which contains minerals but little humus. Deep plant roots grow through the subsoil looking for minerals. Water carries clay and minerals down into the subsoil where it builds up. A layer with large pieces of broken rock is below the subsoil. These four layers rest on a layer of solid rock, called bedrock.

Soil contains parts from all four Earth systems, from tiny pieces of rocks and minerals, to humus. A typical forest soil has four layers.

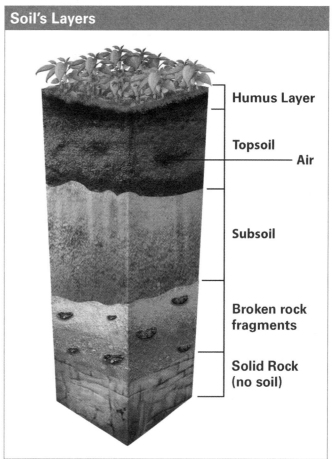

Soil's Layers

Humus Layer

Topsoil

Air

Subsoil

Broken rock fragments

Solid Rock (no soil)

Different environments on Earth have different types of soil. Write a brief sentence explaining how the soil got that way.

Environment	Soil Type	How Soil Got This Way
	Forest soil has a lot of humus, or decayed plant and animal matter.	
	Desert soil has a lot of minerals but little humus.	
	Soil in mountains is very rocky.	

Show What You Know

Draw a picture of what you see outside.
- Label things in your picture that are part of the atmosphere, biosphere, geosphere, and hydrosphere.
- Identify two places in your picture where two of Earth's systems interact. Describe how the systems are interacting.

Making Sense of the Phenomenon

Let's revisit the phenomenon: *Lakes drain during years of low rain.*

Think about:
- Which of Earth's systems is the lake a part of?
- Which other Earth systems can you see here?

Use your findings from the investigation to answer this question: *How do Earth's four systems work together?*

Claim	
Evidence	
Reasoning	

☑ Go back to page 4 and fill out the unit checkpoint for this lesson.

Lesson 2

How Do Earth's Systems Produce Weather and Climate?

Observing Phenomena

Discuss: What does the sky look like when it is raining?

Observe this phenomenon: *The fog in this valley stays low to the ground.*

See It! From a cozy place, watch a rain storm! The rainwater is part of the hydrosphere, but where is the rain coming from? Is the sky clear or cloudy on a rainy day?

Think of what you already know about how Earth's systems produce weather and climate. Write questions you have.

Creating a Weather Report

Now you will create a weather report to explain how Earth's systems interact to produce the weather or climate on the placard.

Your teacher assigned you a type of weather or climate to analyze. Create a diagram that shows how Earth's systems interact to produce the weather or climate.

Make sure to:
• title the diagram.
• label Earth's systems that interact to produce the weather or climate.

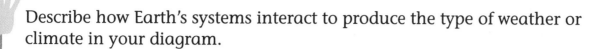

Describe how Earth's systems interact to produce the type of weather or climate in your diagram.

Use your diagram and description of how Earth's systems interact to create a weather report. Your report should:
- be no longer than 1 minute.
- include a diagram of the weather or climate.
- describe how Earth's systems interact to produce the weather or climate. Make sure to name the systems that interact.

Presenting a Weather Report

It's time to present your reports!

Present your weather or climate when your group is called on.

Vocabulary

Match the term to the correct definition.

Word Bank

climate   prevailing wind   air mass   weather

_____ 1. A large quantity of air that has similar temperature, moisture, and pressure all through it.

_____ 2. The general weather of a place over a long period of time, such as many years.

_____ 3. Wind that usually blows more often from one direction than from any other direction.

_____ 4. The condition of the atmosphere at a place for a short period of time, such as a few hours or days.

My Science Concepts

Reflect on your understanding. Draw an X along each line.

The atmosphere is always involved in producing weather and climate.

still learning   **know it**

Precipitation is caused by interactions between the atmosphere and hydrosphere. When rain, snow, or hail fall from the clouds and hit the ground, it interacts with the geosphere.

still learning   **know it**

Some places in the north have a cool climate. If the hydrosphere did not take warm currents to the shores of those places and warm up the air in the atmosphere, the climate would be much cooler.

still learning   **know it**

1. Earth's Systems Interact

Your baseball game is postponed because of a summer storm of dense black clouds, rain, streaks of lightning, and loud thunder. A thunderstorm can result when Earth's atmosphere and hydrosphere interact.

Earth's four systems interact with each other. They can even interact with each other to produce the weather that you experience. For example, water vapor is part of both the hydrosphere and atmosphere. When the water vapor high in the atmosphere cools and changes to water droplets, clouds form. Wind results when heat from Earth's surface, the geosphere, warms the air in the atmosphere unevenly. Moving water in a river wears away soil and rock to change the shape of the surface of the Earth. Plants and animals in the biosphere exchange gases with the atmosphere when they make food or breathe.

All of the weather and climate you experience are produced by interactions of Earth's systems. **Weather** is the condition of the atmosphere at a place for a short period of time, such as a few hours or days. **Climate** is the general weather of a place over a long period of time, such as many years. A thunderstorm is a kind of weather that resulted from an interaction between parts of the hydrosphere and parts of the atmosphere. Warm summers and cold winters may be the climate in your area. Climate is affected by many things, such as how close an area is to oceans and mountain ranges.

A thunderstorm is a kind of severe weather. It results from interactions between parts of Earth's hydrosphere and its atmosphere.

Write a paragraph describing the weather today. Which Earth systems are interacting?

Write a paragraph describing the climate of the area you live in. Can you think of any reasons that your climate is like this?

2. Air Pressure and Temperature

The weather is always changing. Many factors in the atmosphere affect the weather, such as the air pressure and temperature in that area.

Air Pressure

You have learned that air is made of tiny invisible pieces of matter. The matter in air pushes down on the surface of Earth. **Air pressure**, or how much air pushes on any surface, affects the weather. So, air pressure is affected by the amount of water vapor in the air. Water vapor makes air moist and lowers the air pressure. When the air pressure changes, the weather usually does too. Rising air pressure usually means dry weather is nearing while falling air pressure usually means wet weather is approaching.

Temperature

Temperature also affects the weather. The particles of matter in cold air are packed more tightly together. So, cold air weighs more than the same volume of warm air. The cold air then sinks down and pushes the warm air up. Earth's surface, its hydrosphere and geosphere, affects the temperature of the air. During the day, the sun warms the land and oceans. Energy transfers from the land and water, and heats the atmosphere. At night, the air cools.

The amount of change in air temperature also affects an area's climate. Land areas near a warm ocean tend to be warmer than inland areas, and land near cold oceans tends to be colder. Even the color of a land area affects air temperature. Dark rocks absorb more of the sun's energy. White snow and ice reflect most of the energy off Earth's surface. Some of the reflected energy is absorbed by gases in the atmosphere.

During the day, the sun warms the land and oceans. Heat is transferred to the atmosphere and warms the air.

Organisms in the biosphere can affect air and land temperatures. You may have noticed that the air in a room warms up when it is crowded with people. Some of the body heat of animals is transferred to air.

Air Masses, Fronts, and Weather Maps

An **air mass** is a large quantity of air that has similar temperature and moisture all through it. A single air mass may be large enough to cover several states.

Air in the atmosphere is always moving. As it moves, energy and water vapor are transferred from the land and water to the air, or from the air back down to the land or oceans. The place where one air mass meets another air mass is called a *front*. Weather changes at fronts. Suppose a cold air mass is moving toward a warm air mass. This is called a cold front. Cold air cannot contain as much water vapor as warm air can. So there might be rain or thunderstorms at the front. This can be followed by cool dry weather as the cold dry air mass replaces the warm air mass in the area.

What happens when a moving warm air mass meets a cold air mass? This is called a warm front. The incoming moist air pushes over the top of the cold air mass. The rising warm air cools off, often producing rain. After the warm front passes by, the weather gets warmer and clearer.

Air masses and fronts can be shown on a weather map. The numbers and symbols on a weather map identify the air masses. Blue triangles identify cold fronts and red semicircles identify warm fronts.

A cold front forms when a moving cold air mass runs into a warm air mass, pushing the warm air upward. A warm front forms when a moving warm air mass runs into a cold air mass. The warm air slides over the cold air. A cold front is shown by blue triangles while a warm front is shown by red semicircles.

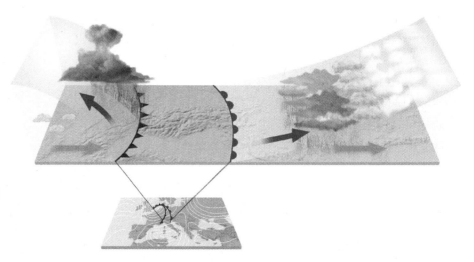

What is it called when a moving warm air mass moves into a stationary cold air mass?

Draw a diagram showing what happens when a moving warm air mass moves into a stationary cold air mass.

Write a sentence explaining your diagram.

Cold air does not hold water vapor very well. What predictions can you make about the weather as the warm air mass moves in?

3. Clouds and Precipitation

When you look up at the sky, you might see white puffy clouds or black storm clouds in the atmosphere. Can you make predictions about the weather by observing the different shapes and colors of clouds?

Formation of Clouds

Clouds form when Earth's hydrosphere interacts with the atmosphere. The air cools and water vapor condenses, turning into liquid water droplets. These droplets are a cloud. If the air is cold enough, as it is high in the upper atmosphere, the droplets freeze to form ice crystals.

One way clouds can form is when a warm, moist air mass meets a mountain and rises. Mountains located near oceans often have clouds over them because the air picks up moisture over the ocean. Clouds also form when warm, moist air is pushed upward by a front.

Kinds of Clouds

The kind of clouds in the sky can tell you what the weather is going to be. *Cirrus clouds* are feathery clouds that form high in the atmosphere and are made of ice crystals. Cirrus clouds usually mean good weather. *Cumulus clouds* are fluffy and white with flat bottoms that form lower in the atmosphere than cirrus clouds do. Cumulus clouds mean good weather. *Stratus clouds* are flat, gray, and layered. They are the lowest clouds in the sky and usually cover the whole sky. They might bring rain and drizzle. *Cumulonimbus clouds* are storm clouds. They are dark and heavy and usually mean that a strong storm is approaching.

Clouds form when Earth's hydrosphere interacts with the atmosphere. Cirrus clouds are made of ice crystals that form high in the atmosphere.

Cumulonimbus clouds are dark and heavy in the atmosphere. When you see cumulonimbus clouds you know that a strong storm is coming.

Formation of Precipitation

Remember that precipitation is water that lands on the Earth's surface, in other words, an interaction between the hydrosphere and the geosphere. After clouds have formed, the water droplets in the clouds grow larger and heavier as water vapor continues to condense. When the droplets get too large and heavy, they fall to Earth.

Kinds of Precipitation

There are several kinds of precipitation. They vary according to the size of the droplet and the state of water. Rain is liquid water drops that fall from clouds. Some raindrops are quite large, whereas others are small. Small raindrops are sometimes called *drizzle*. Snow is frozen water that is made of ice crystals called snowflakes. Each snowflake has six sides or points and has a different shape. Sleet is small drops of ice that are formed when falling snow melts and then freezes again. *Hail* is made of pellets or chunks of ice. A hailstone forms when a frozen raindrop is pushed upward in rising warm air. More water freezes on it, and makes it a larger pellet of ice. When the hailstone gets too heavy to be suspended, it falls to the ground.

These hailstones are the same size as blueberries. Hailstones can be as large as baseballs.

Fog

Fog is small liquid water drops that lie close to the ground. You can think of fog as being a cloud that forms at ground level. Fog and clouds are made of water droplets. However, clouds form from rising and cooling moist air. Fog forms when air moves over a cool surface. Thus, fog forms when the atmosphere and geosphere interact. Clouds form when the atmosphere and hydrosphere interact.

Observe the clouds in your area. Draw a picture of what they look like.

What predictions can you make about the weather from what the clouds look like?

What predictions about precipitation can you make from the temperature of your area?

Which Earth systems interact to produce clouds and precipitation?

4. Ocean Currents

In 1992, 12 boxes of bathtub toys were washed overboard into the hydrosphere from a ship in the Pacific Ocean during a storm. Within 10 months, colorful blue turtles, yellow ducks, red beavers, and green frogs began to appear on beaches throughout the geosphere. Some of these toys were washed up in Australia and Alaska. Others landed in western South America. Scientists think the last of the toys will wash up in England.

The Ocean in Motion

The water in Earth's ocean is constantly moving. Ocean water that flows from one place to another is called an ocean current. An ocean *current* is like a river of water within an ocean. Different currents have different properties and effects. Some flow near the ocean's surface. Others are much deeper. Some currents consist of cold water, but others are warm. The floating bathtub toys traveled around the world on surface ocean currents. Earth's ocean currents can be mapped. Maps show that huge systems of surface currents move water in a circular pattern. Each system is made up of several main currents.

How Surface Currents Form

Surface currents result from interactions between the hydrosphere and the atmosphere. Wind is moving air. It pushes on the ocean's surface and moves the water in a horizontal direction. Some surface currents last for only a short time in a small area. Many, however, are long lasting. They cover large parts of the ocean.

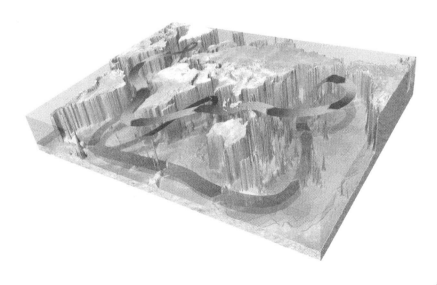

Earth's ocean current systems move and mix water in a circular pattern. Within each circle are several currents.

How Deep Water Currents Form

Deep water currents are caused partly by temperature differences in the ocean. Cold water is usually heavier than warm water. When the cold water meets warmer water, the cold water sinks. This causes a vertical current.

The Effects of Ocean Currents

Surface currents have an important effect on climates. The Gulf Stream, one of the best known warm currents, flows along the East Coast of the United States. Then it turns east and crosses the Atlantic Ocean, when it is called the North Atlantic Current. The Gulf Stream carries warm water north from the Gulf of Mexico. As it flows, it transfers heat from warmer areas of Earth to cooler areas. England and much of northwestern Europe would be much colder in winter if it were not for this current.

Cold currents start near the poles. They carry cold water toward the equator. California would be warmer if not for the cold current from the Arctic. Western Africa would be warmer if not for the cold current from the Antarctic. Thus, ocean currents play an important role in determining the climate of an area.

Ocean water contains tiny organisms that are food for fish and other ocean animals. Ocean currents carry this food from one part of the ocean to another. This interaction of the hydrosphere and the biosphere provides food for marine animals so they can live in different areas of the ocean.

Palm trees grow in southern England because warm ocean currents warm the climate. Southern England is almost as far north as Alaska, which is not warmed by ocean currents.

Which Earth system does the ocean interact with to form surface currents?

How does the ocean form deep water currents?

How do ocean currents affect climates?

How do ocean currents affect the fish in the biosphere?

5. Prevailing Winds

Why does a plane trip from New York to London take less time than a similar plane trip from London to New York? Winds moving from west to east through the atmosphere help push the plane east.

Wind is moving air. Air moves because Earth's surface is heated unevenly. Land heats up faster than water does so the air above a beach becomes warmer than the air above the water. The warm air rises and is replaced by cooler air that moves in from above the water. This is a sea breeze.

Winds also form high in the atmosphere. When the wind blows more often from one direction than from any other, it is called a **prevailing wind**. In the United States, prevailing winds generally blow from west to east. The winds carry clouds and air masses with them. Weather seen in the west on Monday will usually be seen in the east later in the week.

The pattern and direction of wind currents in the lower atmosphere is very similar to that of the surface ocean currents in the hydrosphere. You may have heard of a jet stream. Jet streams are fast flowing worldwide air currents similar to ocean currents. Most blow from west to east. A jet stream usually flows through the middle of the United States, separating the cooler air in the north from the warmer air in the south, but it can move. In winter, it may dip to the south. Then it brings cold air to the southern part of the country. If it moves north in summer, it can bring a heat wave north. A jet stream can produce low-pressure areas and stormy weather in the air underneath it.

Orange growers sometimes lose their crop when the jet stream dips south and brings freezing weather to Florida.

© Teachers' Curriculum Institute

Draw arrows and label them to show how a breeze can form. Include these terms: **warm air** and **cool air**.

Draw and label how jet streams and prevailing winds can carry clouds and air masses.

Show What You Know

Record observations of what the weather is like where you are.
Fill in the information below.

Location _____ Date _____

Temperature: _____

Precipitation: _____

Weather description _____

Complete the paragraph below to explain how two of Earth's systems interact to produce the weather where you are.

The weather where I am is _____

Two of Earth's systems that interact to produce this weather are _____

_____

These systems work together to produce the weather by _____

_____

_____

_____

_____

Making Sense of the Phenomenon

Let's revisit the phenomenon: *The fog in this valley stays low to the ground.*

Think about:
- Which Earth system is this fog a part of?
- Which Earth systems interact to create this fog?

Use your findings from the investigation to answer this question:
How do Earth systems interact to change the weather and climate?

Claim	
Evidence	
Reasoning	

☑ Go back to page 4 and fill out the unit checkpoint for this lesson.

Lesson 3

How Do Earth's Systems Change Earth's Surface?

Observing Phenomena

Discuss: Have you been to a rocky beach? Why do you think the rocks are the shape they are?

Observe this phenomenon: *These rocks are oddly shaped.*

Make It! On a windy day, leave a small pile of dirt or sand outside. What happens to the pile after a few minutes? A few hours?

Think of what you already know about how Earth's systems change Earth's surface. Write questions you have.

Exploring Changes to Earth's Surface

Your teacher will hand your group a slip of paper that describes a change to Earth's surface. Read the description with your group.

You will create an act-it-out to model how Earth's systems interacted to produce your change to the land.

Creating Your Act-It-Out

You were assigned an important role in creating your act-it-out. Find your role and complete your responsibilities. Then, come together as a group and rehearse your act-it-out.

Researcher:

In the table below, describe how each of Earth's systems is involved with producing the landform you were assigned. If it does not affect the creation of the landform, you can leave the square blank.

Atmosphere	Biosphere
Hydrosphere	Geosphere

Write an explanation of how Earth's systems interact to change Earth's surface to produce the landform you were assigned.

Playwright:

Write a script for an act-it-out that describes and models how Earth's systems interact to produce the landform you were assigned.

- Make sure everyone in your group plays one of Earth's four systems.
- Every system should explain if it has a role in changing Earth's surface to produce the landform and how it produces it.

Presenting Your Act-It-Out

As each group presents their skit, fill out the table. Briefly describe what role each Earth system has in producing the land form. If the system does not play a role in producing the landform, leave that space blank.

Landform	Atmosphere	Biosphere	Geosphere	Hydrosphere
Barrier islands				
Continental islands				
Coral islands				
River delta				
Rock formations				
Sea caves				
Soil				
Volcanic islands				

Vocabulary

Match the term to the correct definition.

Word Bank

deposition   erosion   landform   weathering

_____ 1. A natural structure on Earth's surface.

_____ 2. The dropping of weathered material in one place.

_____ 3. The breaking down of rock by interactions with Earth's systems.

_____ 4. The loosening and carrying away of weathered material from one place to another.

My Science Concepts

Reflect on your understanding. Draw an X along each line.

The atmosphere, hydrosphere, and biosphere break down, carry away, and drop parts of the geosphere. Pieces of rock are eroded when wind or water carry them away. When the pieces are dropped, they are deposited.

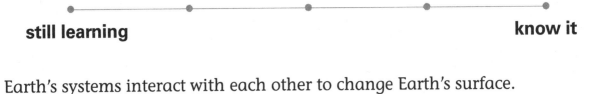

still learning   **know it**

Earth's systems interact with each other to change Earth's surface.

still learning   **know it**

Creating an act-it-out helps you articulate how Earth's systems interact to change Earth's surface. These changes can happen over a long period of time. Modeling these interactions can help you understand how this happens in a shorter amount of time.

still learning   **know it**

1. Earth's Surface Changes

What changes occur to Earth's surface when systems interact? Millions of years ago, a glacier dug out a deep valley in the area where Yosemite National Park is today. A river flowing nearby met a steep wall of the valley and plunged over it. Bridalveil Falls was created in the river from an interaction between the hydrosphere and the geosphere.

Earth's surface is always changing. Many of these changes occur because Earth's systems interact. Mountains are worn down by water and wind, changing their shape and forming new soil over time. Glaciers in the hydrosphere pick up rocks and soil in the geosphere and carry them far away. The rock and soil forms hills when they stop moving. The hydrosphere's flowing water carries away soil from river banks and beaches in the geosphere. The water and wind deposits the soil in new places to build new beaches, sand dunes, and other natural structures. These natural structures on Earth's surface are called **landforms**.

Plants and animals in the biosphere change Earth's surface in many ways. Plant roots break up rock, part of the geosphere. Decaying plant matter in the biosphere helps to form new soil for the geosphere. People remove soil when they mine minerals. They build dams, creating new bodies of water where forests and hills once stood.

As Earth's surface changes, the other systems change. Weather and climate may change if a new mountain range in the geosphere causes more or fewer clouds to form. Animals may leave an area if weather patterns change.

Bridalveil Falls in Yosemite National Park was created when a glacier gouged out the valley and changed the landscape.

Choose two of the four landforms shown here. Write a brief explanation for each on how you think they each might have been formed. Include at least two Earth systems in each of your explanations: **atmosphere,** **biosphere, hydrosphere,** and **geosphere.**

Waterfall

Beach

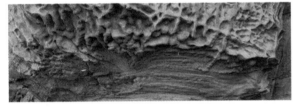

Sandstone rock formation

Sand dune

Landform:

Landform:

2. How Islands Form

When you think of an island, you may picture a sandy area with a palm tree surrounded by water. Islands are surrounded by water, but they are not always sandy nor have palm trees. Manhattan is an island covered with tall buildings. There are four kinds of islands and they form in different ways.

Continental Islands

A continental island forms when a chunk of land breaks off from a continent, a major land mass. This piece becomes an island. Continental islands form when the hydrosphere and geosphere interact. Greenland, Earth's largest island, formed in this way millions of years ago.

Continental islands can also form when changes in sea level occur. Thousands of years ago, the northern part of Earth was covered with glaciers. Sea level was much lower than it is today. As the Earth warmed, the glaciers began to melt and the sea level rose. The ocean covered some of the low-lying land, creating continental islands. Great Britain was formed in this way. It was once part of mainland Europe.

Volcanic Islands

There are volcanoes on the ocean floor that erupt to release lava that hardens on the ocean's floor. When a volcano erupts over and over again, layers of lava build up until it appears above water and an island is formed. Lava is molten rock from the Earth's mantle that reaches the Earth's surface. Thus, volcanic islands form when the hydrosphere and geosphere interact. Volcanic islands are made of igneous rock. The islands of Hawaii are volcanic islands.

The Hawaiian Islands were formed when underwater volcanos erupted. Layers of lava stacked on top of each other until they formed the islands you see today.

Barrier Islands

Barrier islands also form by the interaction of the hydrosphere and the geosphere. They are narrow islands that form near coastlines when ocean currents deposit piles of sand and other sediments. Eventually the piles rise above the water's surface and become islands. Barrier islands act to protect the coast from ocean waves and wind during storms. Barrier islands can also form in rivers when sediments are deposited.

Barrier islands can also form when glaciers melt. The sea level rises and floods areas around sand dunes, creating sandy islands. The Outer Banks along the southeastern coast of the United States formed this way. Glaciers pick up rock, soil, and gravel as they move over land. This material piles up when a glacier melts. The piles are surrounded by water as sea level rises.

Coral Islands

Coral islands form when the biosphere and hydrosphere interact. Tiny sea animals called corals live in warm ocean water. Corals produce hard skeletons made of limestone outside their bodies that are similar to the shells of clams. Corals live in large colonies that contain thousands of organisms. A coral colony forms a huge limestone reef on the ocean floor. As corals reproduce, the reef grows larger. It grows upward in layers until it breaks through the water's surface. A coral island is formed. Coral islands are made mostly of limestone but sand and other kinds of rock may be included. The islands of the Bahamas and the Florida Keys in the Atlantic Ocean are coral islands.

The Florida Keys are coral islands. You can see much of the submerged coral reef through the clear water.

Explain how different Earth systems can interact to form islands.

- Include at least two of these terms: **biosphere**, **hydrosphere**, and **geosphere**.
- Include direct quotes from the Student Text.

3. Weathering, Erosion, and Deposition

What would happen if you hit two rocks against each other above a table? You might find crumbs of rock on the table. If you hit the rocks hard and long enough, you might even break off a large piece. Rock on the Earth's surface can also crumble and change shape.

Weathering

Earth's surface and its landforms are always changing. Many changes take millions of years. **Weathering** is one of these slow changes. Weathering is the breaking down of rock by interactions with Earth's systems. Wind carries tiny pieces of sand. They hit and rub against rock and wear it away. This is an atmosphere-geosphere interaction. Rain, rivers, and ocean waves also carry tiny pieces of sand. They beat against rock and weather it. Water also seeps into cracks in rock. The cracks expand when the water freezes. Eventually, pieces of rock break off and are carried away. Glaciers gouge out rock when they move over it. These are examples of hydrosphere-geosphere interactions. Plant roots grow into cracks in rocks when the roots search for water. As roots grow larger, they expand the cracks and push apart the rock. This is a biosphere-geosphere interaction.

Weathering changes the shape of landforms. Water moving through the geosphere can slowly hollow out a large area of rock. Some caves form underground when water trickles through soil and rock. Sea caves form when waves weather rock at the base of cliffs. Wind weathered rock in the Sahara Desert in North Africa to create strange rock shapes.

This sea cave was formed by weathering of the rock cliff by ocean waves.

These rock formations were created when wind weathered the soft sandstone rock in the Sahara Desert.

Erosion

Sediments form from weathered rock. What happens to the sediments? **Erosion** is the loosening and carrying away of weathered material from one place to another. Wind in the atmosphere, and water and glaciers in the hydrosphere cause erosion. They pick up the sediments and carry them away. The amount of erosion is greatest when the wind or water moves fast for long periods of time. It took millions of years for water in the Colorado River to weather and erode enough rock to form the Grand Canyon.

Deposition

When moving wind, water, and glaciers in the atmosphere and hydrosphere slow down, they drop the sediments they carry. This process is called deposition. New landforms are created by **deposition**. For example, rivers slow when they approach the ocean. They drop their load of sediments to the river bottom. Eventually, a *river delta* forms. A river delta is a large flat area of land at the mouth of a river. Soil in a river delta is often very rich because the river erodes nutrient-rich soil and deposits the soil at the river delta.

New beaches form where sediments in ocean water are deposited. The shape of a river may change as soil is eroded from the outside of a bend and deposited on the inside of the bend, where water flow is slower and does not push as had on the river bank. Glaciers deposit gravel and rocks at their front and sides before and as they melt and shrink. Ridges of rocks form. The geosphere is shaped by the different Earth systems.

Erosion occurs on the outside of a river bend because water is flowing faster there and rubs harder against the river bank. Deposition occurs on the inside of the bend where water flow is slower and sediments are dropped.

Choose a landform to research. You can use your Student Text or find information on the Internet.

Draw a picture of this landform.

Write a short report on how this landform was formed.

- Include these terms: **weathering**, **erosion**, and **deposition.**
- Include any relevant Earth systems, such as the **atmosphere**, **biosphere**, **hydrosphere** or **geosphere.**

4. How Soil Forms

Imagine that a volcano erupts and covers the soil with lava that hardens into a thick layer of rock. Within a few years, tiny plants begin to grow in cracks in the rock. Plants need soil to grow. How did this soil appear?

This plant has begun to grow in the crack of a rock. A tiny bit of soil has blown into the crack. The plant roots will help make more soil.

Soil contains parts from all four of Earth's systems. It has pieces of rock and minerals, bits of decaying plant and animal matter, water, and air pockets. The pieces of rock and minerals in soil come from sediments, or weathered rock. It is mixed with bits of decaying plant and animal matter from organisms that have died. Water comes from recycled precipitation and runoff from the water cycle. Air comes from the atmosphere. The parts mix together to form soil. Worms and other living things in soil help to mix together the parts of soil. Worms and plant roots make channels in soil. Air enters the soil through these channels.

It can take more than a thousand years for new soil to form. Climate plays a role in soil formation. Warm temperatures and a lot of rain cause soil to form faster because weathering is faster. Soil forms more slowly in cold dry climates where less weathering occurs. Soil accumulates faster on flat land because newly formed topsoil is not carried away quickly. On steep slopes such as mountainsides, new soil may be eroded almost as soon as it forms.

Soil is carried by wind and water to new places, where it is deposited. Small amounts may be deposited in the crack of a rock. If a seed falls into this tiny bit of soil, it begins to grow. The developing roots help to break down the rock. The broken-down rock is added to the soil to make more soil.

Explain how plants can grow in small cracks of rock that do not appear to contain soil.

- Include these terms: **weathering**, **erosion**, and **deposition**.
- Include any relevant Earth systems, such as **atmosphere**, **biosphere**, **geosphere**, or **hydrosphere**.

Show What You Know

Turn an image of a landform into a diagram!

- Find an image of a landform, and paste it below.
- Add arrows and labels to show how Earth's systems interact to produce the landform.
- Underneath the diagram, describe which systems help produce the landform and how they interact to form it. Use the terms **weathering**, **erosion**, or **deposition** if they are involved with producing the landform.

Making Sense of the Phenomenon

Let's revisit the phenomenon: *These rocks are oddly shaped.*

Think about:
- How has Earth's systems affected these rocks?
- Which Earth systems do you think interacted to create these rock formations?

Use your findings from the investigation to answer this question: *How do Earth's systems interact to change Earth's surface?*

Claim	
Evidence	
Reasoning	

☑ Go back to page 4 and fill out the unit checkpoint for this lesson.

Performance Assessment:
Writing an Article on Earth's Systems

As an intern for the EPA, research the hydrosphere and how it affects other Earth's spheres around the state of California. Then, you will use this information to write an article explaining how Earth's spheres interact.

In this Performance Assessment, you will:
- graph the amounts of salt water and fresh water on Earth and where it is located.
- examine a study to determine how Earth's sphere's interact.
- write an article describing Earth's systems and their effects on one another.

Understanding Interactions within the Hydrosphere

Recall that about three-fourths of Earth's surface is covered in water.

This water in the hydrosphere comes into contact with the geosphere, atmosphere, biosphere through the water cycle.

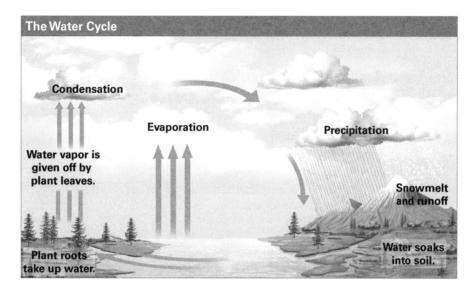

Complete the table by providing an example for how the hydrosphere affects each of the spheres in a normal system. Then, describe how the spheres interact with one another in your example.

	Example of Interaction
Atmosphere	
Biosphere	
Geosphere	

Examining Drought in Tulare County

Some areas experience droughts, or long periods of time in which there is very little or no rainfall.

What kinds of changes will happen in the other spheres if there are changes in the hydrosphere because of a drought? Let's examine this question as we read *Handout: Drought in Tulare County*.

Complete the table by providing an example for how the hydrosphere affects each of the spheres during a drought. Then, describe how the spheres interact with one another in your example.

	Example of Interaction
Atmosphere	
Biosphere	
Geosphere	

Now, draw a model showing one of the examples of an interaction that you listed in the table. Include labels in your models that explain:

- which spheres are interacting.
- how they are interacting.
- what the outcome of this interaction is.

Hydrosphere and Atmosphere Model

Hydrosphere and Biosphere Model

Hydrosphere and Geosphere Model

Writing Your Article

Using the table of information that you created, write a four paragraph article that answers the question: How are Earth's spheres all connected?

Your article should include:
- one paragraph that includes your claim.
- one paragraph for each example that you have.
- one paragraph that explains your reasoning.

Use your graph of the hydrosphere, your models, and evidence from the case study to support your argument.

Lesson 4

How Do Farming and Industry Affect Earth's Systems?

Observing Phenomena

Discuss: How do factories affect global warming?

Observe this phenomenon: *The miner is digging in the earth.*

Make It! Set up a treasure hunt! Hide something in a container of dirt or sand, then have a partner dig to find it. What happens to the dirt or sand as your partner digs?

Think of what you already know about how farming and industry affect Earth's systems. Write questions you have.

Preparing for Mining Chips

In this investigation, you will model the effects the mining industry has on the geosphere by "mining" chocolate chips from chocolate chip cookies. Don't eat the cookie! Use it as a coal mine model instead.

How many chips do you think you can mine in two minutes?

Mining Chips Round One

The work day is about to begin! Prepare to mine chips from the cookie.

Now that you are done, it's time to sell the coal!
How many chips did you mine in two minutes?

Number of chips: _____ Tool(s): _____

Total money earned: $ _____ (number of chips x $1,000)

Ask these questions in your group:
• Whose actual results came closest to their prediction?

• Who made the most money in your group?

• Whose cookie was damaged the most?

• What were these students' mining techniques?

Now examine the effects mining had on your group's cookies. Discuss:
• How did mining damage the cookies? Point to specific parts of the cookies as evidence.

• If the government decides to fine your mine $200 for every piece of land that breaks off, how will your mining techniques change?

What tools and chip extraction strategies will you use in Round 2 to prevent damage to the environment while still extracting as many chips as possible?

Mining Chips Round Two

Another work day is about to begin!

Now that you are done, it's time to sell the coal again!

Record the number of chips you extracted and the number of broken cookie pieces.

Number of chips: _____ Money earned: $ _____ (number of chips x $1,000)

Number of broken pieces: _____ Money fined: $ _____ (broken pieces x $200)

Total money earned: $_____ (money earned – money fined)

Ask these questions in your group:
- Whose actual results came closest to their prediction?

- Who earned the most money in your group?

- Who was fined the least in your group?

- What were these students' mining techniques?

Comparing Results

Compare your results from each round and answer the following questions.

a) Was the cookie environment damaged more in Round 1 or Round 2?

b) How did the $200 fine affect how you mined during Round 2?

c) How would you change your mining technique to make more money next time?

Talk with your group: How do you think mining cookies represents the effects of mining on land?

Vocabulary

Match the term to the correct definition.

Word Bank

pollution   toxic

_____ 1. Anything in the environment that can harm living things or damage natural resources.

_____ 2. Capable of causing injury or death to an organism.

My Science Concepts

Reflect on your understanding. Draw an X along each line.

Farming affects Earth's systems by harvesting crops and using fertilizers that have harmful runoff. Mining carves into the Earth's geosphere and industry machinery releases harmful chemicals into the atmosphere.

still learning   **know it**

The effect of coal mining on Earth's systems can be reduced by fining industries when damage occurs.

still learning   **know it**

Seeing the effects of mining on an actual coal mine takes a lot of money and time. Using a cookie as a model gives you an idea of how mining affects the geosphere and how a fine can change the effects of mining without spending a lot of money and time. A model does not harm the environment.

still learning   **know it**

Although a cookie model can be helpful, it is also inaccurate. The texture of the cookie is different from Earth's rock, and the tools you use to dig cookies are different from actual mining machinery.

still learning   **know it**

1. Humans Affect Earth's Systems

Coal, crude oil, and natural gas are fossil fuels. You have probably heard that Earth's fossil fuels are being used up. People use them much faster than nature can make them. Why are people using so many fossil fuels? What are people doing to prevent them from being used up?

Humans are part of Earth's biosphere. Although the biosphere contains the least amount of matter of Earth's four systems, and humans make up only a small part of it, human activities have a huge effect on Earth's systems.

Most fossil fuels are used in power plants to make electricity. You might use electricity to keep your home warm in winter and cool in summer. People also use electricity to run machinery in factories. People use fossil fuels in cars, buses, and trucks to move people and things from place to place. As more people use fossil fuels to meet their many needs, the supply is being used up. Scientists have estimated that the world's supply of crude oil will be gone between 2050 and 2075 if we continue to use it as quickly as we do now.

Scientists and engineers are working to develop other energy sources. If people use solar, wind, and water power to make some of the electricity people need, then less fossil fuels will be used. If people build cars that run on batteries, they will save more fossil fuels.

There are many ways humans have an impact on Earth's systems. Two ways are through farming and industry.

Humans affect Earth systems whenever they use electricity. More than two-thirds of all fossil fuels, a part of the geosphere, are used to make electricity.

© Teachers' Curriculum Institute

Lesson 4   How Do Farming and Industry Affect Earth's Systems?

85

Humans use fossil fuels every day for many different purposes. List two different ways you use fossil fuels.

I use fossil fuels by riding/driving a car. and I am useing fossil fuels by useing eletricity.

Scientists and engineers are working to develop energy sources other than fossil fuels. What are some other energy sources scientists and engineers are researching?

Some other energy sources are solar water, and wind.

2. Farming Affects Earth's Systems

Humans on Earth need food to survive. Most food comes from large commercial farms that raise fruit, vegetables, and animals used for meat. You may not notice, but farming actually affects Earth's systems.

Farming Affects the Hydrosphere

Most farmers add fertilizers to the soil. Fertilizers provide crops with nutrients. Farmers also spray their crops with *toxic* chemicals that kill weeds and harmful insects. **Toxic** materials are materials that cause injury or death to an organism. But fertilizers and toxic chemicals can also cause *pollution.* **Pollution** is the presence of anything in the environment that can harm living things or damage natural resources.

Large-scale chicken farms can produce thousands of eggs in a small space, but they produce a lot of waste. Wastes often pollute water.

Fertilizers and toxic chemicals cause water pollution when rain washes them from soil. Animal wastes also pollute water. All of these materials seep into streams, rivers, and ground water that provide our drinking water. They may flow into ponds, lakes, and the ocean and harm life there. Fertilizers can also cause huge numbers of green, plantlike organisms called algae to grow. After the algae die, they decay. The decaying process uses up oxygen that is dissolved in the water, causing many fish that need oxygen to die. Sometimes living things that need less oxygen move into the pond or lake. Other times, there is no oxygen, so the pond or lake becomes "dead." A body of water is "dead" when nothing can live there. Parts of the Gulf of Mexico are dead because there is so little oxygen in the water.

Fields of crop plants and the thousands of animals that large-scale farms raise need a lot of water. Farmers must irrigate their fields and provide water for animals. Using this fresh water can cause water shortages for people.

Farming Affects the Atmosphere

Livestock, especially cows, produce a large amount of methane gas in their intestines. This gas is released into the air and affects the composition of the atmosphere. Recall that carbon dioxide and water vapor absorb energy from the warm surface of Earth. Methane does, too. Millions of cows are raised on large-scale farms. The large quantities of methane they produce contribute to global warming.

Farmers that work large farms spray toxic chemicals onto their crops by plane to kill weeds and insects. Occasionally, these chemicals mix with the air and pollute it. The chemicals may spread onto living things that will be harmed by them. Farmers must be very careful not to spray in windy weather or when rain is predicted.

Farming can create pollution in the atmosphere. Crop duster planes spread chemicals and may pollute the atmosphere.

Farming Affects the Geosphere

Rich topsoil is often blown away by wind when it is plowed. Running water can wash away topsoil when crops are planted on sloping land. Some farmers avoid this by contour plowing. They plow the soil in curved bands that follow the shape of the land. By doing this, running water does not wash away topsoil as easily. Contour plowing reduces soil erosion. It helps the soil hold moisture and nutrients. Tree farmers also prevent soil erosion when they cut only a few trees in a forest instead of clear-cutting the entire forest. This method is time-consuming and expensive, but it leaves tree roots that hold soil.

Farmers are developing different methods to reduce farming's effect on Earth's systems. Contour plowing reduces soil erosion and helps the soil hold moisture.

© Teachers' Curriculum Institute

Research how farming affects Earth's systems. Then, using the Student Text and your research, answer this question in a well-written paragraph: *How does farming affect Earth's systems?*

- Include these terms: **atmosphere, biosphere, hydrosphere,** and **geosphere**.
- Include quotes and cite your sources.

Print sources:

Digital sources:

3. Industry Affects Earth's Systems

Humans have built factories to produce the many things they need. Plants and animals lose their homes when factories are built. Earth's systems are also affected in other ways.

Industries Affect the Hydrosphere

For a long time, engineers have found different ways to change the way a river flows to help industries. They have straightened a river, made it deeper, or blocked it with a dam. People might dig out a river to make it straighter and deeper. Large ships can navigate a straight, deep river more easily than a shallow, curved river. Making a river deeper can also prevent flooding. Building a dam produces water power that is used to make needed electricity. Humans also straighten rivers so roads, houses, and factories can be more easily built.

But changing a river causes problems. River water flows faster when curves are straightened. This increases erosion. Removing plants from a river bank also increases erosion. It also affects the animals that live in and near the river. When dams are built, the environment changes. So does the wildlife. Engineers are beginning to remove some dams that are no longer needed. The rivers will slowly recover.

Nuclear power plants produce electricity. They also produce toxic waste and a lot of heat. The plants are usually built near a body of water such as a river or ocean so engineers can use its water to get rid of the heat. This causes the body of water to warm up, killing fish and other animals. Warming up water can be a form of water pollution. Industries also pollute water when they pour toxic chemicals into a body of water.

Industries sometimes release chemicals and other wastes into rivers. The materials pollute the water and harm fish that live there.

Industries Affect the Atmosphere

Industries release harmful chemicals into the air. These chemicals react with sunlight to form smog, a kind of air pollution. Chemicals formed from burning gasoline in vehicles can also form smog. Smog makes breathing difficult. Some chemicals in the air combine to form acids that dissolve in raindrops. This acid rain kills forest trees and creates dead lakes where fish cannot survive. Governments have tried to prevent acid rain by limiting the amount of toxic chemicals released into the air. They have had some success. But acid rain is still a problem.

You may not have thought of space debris as pollution of the atmosphere, but it is. Discarded and old satellite parts and burned-out rocket stages orbit Earth. This space garbage is dangerous because it may collide with working spacecraft.

Industries produce chemicals that can react with sunlight to form smog. This layer of smog over Los Angeles, California, makes breathing difficult. Smog alerts are issued that warn people to stay indoors.

Industries Affect the Geosphere

Factories need raw materials. These materials are often mined from the geosphere. Mining destroys landforms. It creates land, water, and air pollution. It leaves the soil contaminated with harmful chemicals. Factories also cause land pollution because they generate large amounts of wastes and trash. Some are dumped illegally, and others are added to landfills. Many places are running out of space for landfills. Garbage often contains toxic chemicals. Even newspaper ink can be harmful. These chemicals can seep into soil and harm plants and animals.

Research one of these these industries: nuclear power plant, coal power plant, toy factory, or electronics factory.

Use at least one print and one digital source. Write a paragraph on the effect it has on Earth's systems.

- Include the terms: **atmosphere, biosphere, hydrosphere,** and **geosphere.**
- Include quotes and cite your sources.

Print sources:

Digital sources:

4. Scientists Study Pollution

Suppose a community's river has become polluted with a toxic chemical. Fish are dying. Scientists need to find out where the chemical comes from. They need to find out how the chemical harms fish. And they need to find out how the river water can be made safe again.

Scientists collect information by asking questions. Scientific questions can be answered with data that come from doing experiments. For example, scientists may ask and answer "Does the chemical in the river come from the new shoe factory in town?" and "How does the chemical harm fish?" A question that scientists cannot answer scientifically is "Do you believe this factory should be moved to a different location?" This question asks for an opinion and cannot be answered with data from experiments.

When scientists study pollution, they also learn how people can use science to protect the environment. Some of this information comes from books or science journals. Some comes from doing experiments. And some comes from observations in field studies. The information often helps the engineers design a solution to a problem. They may find out that a pipe at the shoe factory has broken and is leaking toxic chemicals. They may just fix the pipe. They may want to design a pipe made of a different material that will not break. They may find out that factory workers are dumping the toxic chemicals into the river on purpose. If so, they may turn the problem over to the police or government officials.

Scientists study the effect of pollution on fish when they learn that harmful chemicals have entered the water. They ask questions that can be answered with data that come from experiments.

Imagine you are a scientist studying pollution in a river.

Come up with three scientific questions you want to investigate. Remember that scientific questions are questions that are answered by data, not opinions.

What are three sources of data you can use to answer your scientific questions?

Show What You Know

Use your notebook and Student Text to complete the speech bubbles below.

Farmer

I help produce food for people every day. But farming can impact Earth's systems. For example . . .

To collect information about the farmer's impact on Earth, I ask scientific questions. One question I would ask is . . .

To find the answer to this question, I will . . .

Scientist

Miner

I mine resources from Earth's crust to make things like sidewalks, spoons, and computers. But mining can impact Earth's systems. For example . . .

To collect information about the miner's impact on Earth, I ask scientific questions. One question I would ask is . . .

To find the answer to this question, I will . . .

Scientist

Making Sense of the Phenomenon

Let's revisit the phenomenon: *The miner is digging in the earth.*

Think about:
- What happens when a miner digs into the earth?
- How does this digging affect Earth's systems?

Use your findings from the investigation to answer this question:
What techniques can we use to minimize the effects of coal mining on Earth?

Claim	
Evidence	
Reasoning	

☑ Go back to page 4 and fill out the unit checkpoint for this lesson.

Lesson 5

How Do People's Everyday Lives Affect Earth's Systems?

Observing Phenomena

Discuss: Think about everything you use on a daily basis. Where do all the materials to make these things originally come from?

Observe this phenomenon: *You can't find books in nature, but they come from Earth's systems.*

See It! | Look around at the objects in your classroom. What kind of materials are they made from? Where do you think these materials come from?

Think of what you already know about where the materials that make up different items come from. Write questions you have.

Our Everyday Lives and Earth's Systems

You affect Earth's system's everyday! In this investigation, you will observe everyday activities. Then, you will discuss how each activity affects Earth's systems with your group.

Your group will get an Earth system chip for every answer you provide that:

- other groups have not shared.
- is stated in this format: This activity affects the _____ because . . . .

Try to collect as many chips as you can!

Playing at Recess and Earth's Systems

Which Earth systems does playing during recess affect? How are they affected?

Doing Homework and Earth's Systems

Which Earth systems does doing homework affect? How are they affected?

Vocabulary

Match the term to the correct definition.

Word Bank

decompose   recycling

_____ 1. Taking materials from old discarded objects and making new objects from them.

_____ 2. When a material is broken down into smaller pieces, perhaps into its basic components.

My Science Concepts

Reflect on your understanding. Draw an X along each line.

Eating breakfast, driving to school, and using electronics are examples of activities that affect Earth's systems. Food for breakfast is from the biosphere, pollution from driving to school affects the atmosphere and hydrosphere, and resources to make electronics are from the geosphere.

still learning   **know it**

Throwing away food scraps in a compost pile, walking to school instead of driving, and taking care of their electronics are some ways people can reduce the impact their everyday activities have on Earth's systems.

still learning   **know it**

By sharing ideas as a class, you can consider ideas that you and your group may not have thought of. You can also add insight to those ideas that those groups may not have thought of, too.

still learning   **know it**

1. Affecting Earth's Systems in the Morning

Your alarm clock goes off. Still sleepy, you climb out of bed and head for the shower. It's a new day, one in which you will interact with Earth's four systems.

The shower helps to wake you up. Shower water is part of the hydrosphere. As you dry yourself, you notice the label on the towel which says the towel is made of cotton, a plant that is part of the biosphere. You brush your teeth, remembering to turn off the water while you brush. As you know, Earth's supply of fresh water is limited because most of it is trapped in ice. It is important to keep this small amount of usable water clean.

You get dressed and notice that your T-shirt is made of cotton and polyester. Polyester and many other fabrics are made from materials in crude oil, part of the geosphere.

You go downstairs for breakfast, and you smell toast as you walk into the kitchen. The smell of toast is caused by substances in the toast that float from the toast to your nose through air, part of the atmosphere. Toast is made from wheat, another plant in the biosphere. The toaster gets its power from electricity, which is produced by burning fossil fuels from the geosphere. The glass that holds your milk was made from sand. So many of the products you use are from the geosphere. You throw your trash in a trash can so it will be taken to a landfill to decompose. When a material **decomposes** it is broken down into smaller pieces.

You grab your books and run to catch the bus. You have hardly left your house, but you have already interacted many times with Earth's four systems.

Cotton fabric is made from plants in the biosphere. Cotton is a natural product that you might use every day.

Describe what you did this morning when you woke up and got ready for school.

Which Earth systems did you interact with and how?

2. Affecting Earth's Systems at School

As you travel and spend your day in school, you continue to interact with Earth's systems.

In the School Bus

The bus you take to school is made mostly of steel. Steel is a mixture of several metals, such as iron, all of which are mined from Earth's crust. The plastic steering wheel and seats, rubber tires, and glass windows—in fact, most of the bus— are made from parts of the geosphere. The bus is powered by gasoline, a fossil fuel. You know that exhaust gases from buses can pollute the atmosphere and cause smog.

In the Classroom

Books are made of paper which comes from trees. You think about the forest and hope some of the trees remain. You also hope that new trees were planted to replace those that were cut down. Pencils are made from trees, too. The black core of a pencil is made of graphite, a mineral from Earth's crust. Ballpoint pens are made of plastic, a material made from crude oil. The ink in your pen is a complex mixture of many chemicals, and the color in ink often comes from ground-up minerals.

Think of the many school buses that each school needs. Almost all of these buses use fuel from the geosphere and produce air pollution.

In the Cafeteria

It is lunchtime and you head for the cafeteria. All of the food you eat comes from the biosphere, either plants or animals. As you learned, farming can deplete soil and pollute the atmosphere, hydrosphere, and biosphere. Animal wastes can pollute our water sources.

The Obstacle Course in Gym Class

In gym class, you run through an obstacle course made of cones. The cones are made of rubber manufactured from fossil fuels. The wood floor is made from trees. You breathe faster as you run. Breathing takes oxygen from the air and releases carbon dioxide and water vapor. You do not have to worry about using up all the oxygen in the atmosphere because plants produce oxygen when they make food.

As you run, you begin to sweat. Sweat is mostly water which evaporates from your skin and enters the air as water vapor. Your sweat may someday become part of a cloud. Remember that water is cycled among Earth's systems in the water cycle.

The Playground

After school, you and your friends go to a playground. Some of the equipment is made of metal, which comes from Earth's crust, part of the geosphere. Other equipment is made of wood, which comes from trees, part of the biosphere. Plastic parts are often made in factories from oil from the geosphere. If the park has a fountain, the fresh water is part of the hydrosphere.

Some of the equipment from a playground comes from Earth's systems. Climbing bars are made of metal. Metal comes from Earth's crust, part of the geosphere.

Some of your friends use chalk to draw on the sidewalk. Chalk is a sedimentary rock that is taken from Earth's crust, part of the geosphere. Other friends start a baseball game, using a wood bat and a leather ball. Both are part of the biosphere. The wood is from trees and the leather is from animals.

Come up with two ways you affect Earth's systems at school. Be clear and specific about which Earth system you are affecting, and how. Include these terms: **atmosphere, biosphere, hydrosphere,** and **geosphere.**

3. Affecting Earth's Systems at Home

When you get home from school, you relax before starting your homework. You may watch television for a while, call friends, or play a game on the Internet.

A friend may call you before dinner about going for a bike ride. Bikes are made in factories. The raw materials in a bike are mined from Earth's crust. Factories that manufacture the millions of bikes that are sold every year can affect the environment if they cause pollution.

Electronic devices such as television sets, cell phones, and computers are made in factories. They contain hundreds of materials that are taken from Earth's crust. Some metals, such as the cadmium found in computer batteries, are rare. Mercury and lead are toxic. Mercury is found in some light bulbs and flat screen monitors. Plastic cases made from fossil fuels can also be harmful. Factories that make these electronic devices must be careful not to cause pollution.

When electronic devices are discarded, they end up in landfills. The plastic cases do not decompose easily, so they become land pollution. Cathode ray tubes in old television sets and computer monitors contain lead. These devices should not be thrown out with trash. They must be discarded properly so they do not pollute the land and water. Some of their parts can also be recycled. **Recycling** is taking materials from old discarded objects and making new objects from them. You will read about recycling in the next lesson.

Old computers and other electronic devices can pollute land, water, and the atmosphere when they are not disposed of properly.

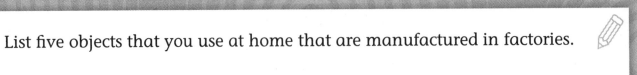

List five objects that you use at home that are manufactured in factories.

What are some ways manufacturing objects in factories impact Earth systems?

Show What You Know

Interview a family member or teacher about an everyday activity.

- Choose an everyday activity. Record it below.
- Come up with three questions. Two questions should be about how the activity impacts Earth's systems. One question should be about how people can reduce the activity's impact. Record the questions below.
- Ask a family member or teacher the questions you came up with. Record who you talked to and their answers below.

Everyday activity: _____

Who I interviewed: _____

Question 1:

Answer 1:

Question 2:

Answer 2:

Question 3:

Answer 3:

Making Sense of the Phenomenon

Let's revisit the phenomenon: *You can't find books in nature, but they come from Earth's systems.*

Think about:
- What are books made of?
- Where do the materials that make books come from?

Use your findings from the investigation to answer this question: *What impact does creating books have on Earth's systems?*

Claim	
Evidence	
Reasoning	

☑ Go back to page 4 and fill out the unit checkpoint for this lesson.

⦿ Lesson 6

What Can People Do to Protect Earth's Systems?

Observing Phenomena

Discuss: What are some of the main causes of pollution that you can think of?

Observe this phenomenon: *Because of pollution, you can't drink this water.*

See It! | On a rainy day, head to your school's parking lot. Why shouldn't you drink the puddles on the ground?

Think of what you already know about pollution and its negative consequences. Write questions you have.

Defining the Problem

You and your group will brainstorm designs for a water filter that removes more dirt from the water.

Define the problem your group is trying to solve. What are your criteria for a successful water filter? What are your constraints?

Criteria:

Constraints:

Designing and Testing Your Filter

With your group, examine the materials you can use. Note the properties of each material. Discuss which properties will be good for which part of the filter.

Decide which materials you will use and what order you will put them into your filter. Once you finish with your design, build your filter!

Draw a diagram of your group's design. Label the materials and explain how each material will help filter water.

Let's test our filters!

Follow these instructions:
- Measure 150 mL of water in a beaker. Get two spoons of soil from your teacher and mix it with the water.
- Hold your filter over a beaker in a plastic tub.
- Twist off the cap.
- Pour the dirty water into the filter.

Observe the effects of your filter on dirty water.
- Describe the water before it was filtered.

- Describe the water after it was filtered.

- Identify the criteria that were met and those that were not.

Creating and Testing Your New Design

Discuss how you can improve your design with your group.

- Which materials worked well? Which ones did not?
- Are there any materials you want to add or remove?
- How can you change the design to meet the criteria better?

Now build your new design!

Draw a diagram of your new water filter design and label the materials. Describe the changes you made to the filter.

Let's test our filters!

To make it a fair test, follow the same instructions you followed for your first design:

- Measure 150 mL of water in a beaker. Get two spoons of soil from your teacher and mix it with the water.
- Hold your filter over a beaker in a plastic tub.
- Twist off the cap.
- Pour the dirty water into the filter.

Observe the effects of your newly designed filter on dirty water.
- Describe the water before it was filtered.

- Describe the water after it was filtered.

- Identify the criteria that were met and those that were not.

Vocabulary

Match the term to the correct definition.

Word Bank

conservation   scrubbers   wildlife refuge

_____ 1. The wise use of material and energy resources.

_____ 2. An area of land or of land and water that is set aside for the protection of wildlife.

_____ 3. Devices that remove dirt and harmful pollutants from smoke produced by burning high-sulfur coal.

My Science Concepts

Reflect on your understanding. Draw an X along each line.

The engineering design process involves clearly defining a problem, designing solutions and weighing their advantages and disadvantages, building and testing the solution, and improving the solution based on tests.

still learning   **know it**

Engineers, individuals, and communities come up with ideas to protect Earth. For example, engineers may develop filters to clean water. Individuals may choose to ride bicycles or walk instead of using a car. Communities may start recycling programs to reduce pollution in landfills.

still learning   **know it**

Conducting research on how human activities affect Earth can help people understand how they interact with Earth's systems. People affect Earth and cause changes to its systems. But if they understand how their activities affect Earth, then they can also come up with ideas to protect Earth.

still learning   **know it**

1. Protecting Earth's Systems Is Important

The dinosaur species *Tyrannosaurus rex (T. rex)* disappeared because Earth's systems changed. Many people think the systems changed because an asteroid hit the earth. Whatever happened 65 million years ago, *T. rex* could no longer meet its needs and went extinct. Can changes in Earth's systems today cause species to become extinct?

Changes to the environment occur all the time, and species naturally become extinct from these changes. Many scientists worry that unbalancing Earth's systems more will lead to many more species becoming extinct. It is important to keep Earth's systems relatively stable so that organisms in the biosphere, such as humans, can meet their needs.

Systems interact with each other, and changes in one system usually cause changes in other systems. If the climate in an area gets colder and drier, less water may evaporate. Fewer clouds may form in the atmosphere, and the amount of rainfall may decrease further. A land area that was once a forest may become a desert.

If the atmosphere warms, glaciers and polar ice may melt and raise the sea level. Low-lying land may be flooded with sea water. The fresh water may become too salty for plants and animals. Some of these organisms may become extinct.

Burning fossil fuels pollutes the atmosphere with carbon dioxide that warms Earth. Polar bears live near the North Pole and hunt for seals from floating pieces of ice, called ice floes. As the Arctic ice warms and melts, fewer ice floes are left. Fewer polar bears can leave land to hunt for seals, and some may starve. Polar bears may soon become extinct.

The fate of polar bears in the Arctic depends on how much the atmosphere warms Earth. More carbon dioxide leads to the earth warming faster.

Complete the flowchart. Explain how burning fossil fuels can result in polar bears being unable to meet their needs and becoming extinct. Be specific!

Burning fossil fuels . . . polots the air in the atmoshere and cases the earth to warm up!

Burning fossil fuels

. . . leads to more global warming and higher temperatures.

Warmer temperatures on Earth lead to . . . the north and south pols ise melting

The ice melting in the north and south pols

. . . means that polar bears are unable to hunt for food.

Polar bears are unable to meet their needs and become extinct.

2. Individuals Can Protect Earth's Systems

The increase in carbon dioxide in the atmosphere is the result of millions of actions. Every time you turn on a light, you use electricity. Generating electricity with fossil fuels produces carbon dioxide. Every time you ride in a car or bus to the library instead of riding your bike or walking, the car or bus's engine produces carbon dioxide. Every day you make choices that affect Earth's systems.

One part of protecting Earth's systems is **conservation**. Conservation is the wise use of material and energy resources so we do not run out of them. You can help conserve these resources by reducing the amount of pollution you produce. When you help wash the family car, what do you do with the soapy water? If you pour it into a storm drain in the street, it will end up polluting a nearby stream or river in the hydrosphere. Instead, your family can take the car to a self-service car wash where the runoff water is collected to be cleaned for reuse at the community's water treatment plant. This keeps the soapy water from polluting a stream or river.

You can also reduce the amount of air pollution you produce by making choices that do not produce air pollution. Riding your bike instead of a car is a healthy choice that produces less air pollution.

What happens to an object after you throw it in the trash? Trash is taken to a landfill to decompose. A landfill is a large area where trash is piled up and buried. Some materials such as plastic do not decompose easily. When they are taken to a landfill, they stay on the ground and produce pollution.

Some objects in a landfill do not decompose easily, and the landfill fills up with pollution. You can conserve resources by reducing the amount of pollution you produce.

Soapy water is a form of water pollution. You can help conserve water resources by washing the family car in a self-service lot, where the soapy water will be collected to be cleaned in a water treatment plant.

Many of the materials that you use in daily life, such as plastics and metals, do not decompose easily. You can help conserve material and energy resources to protect Earth systems by following the three Rs: reduce, reuse, and recycle.

You can reduce how much materials you use in many different ways. When you walk or ride your bike instead of taking a car or bus, you reduce the amount of fossil fuels you are using. Turning off water when you brush your teeth and taking shorter showers reduce your use of fresh water. Buying fruits and vegetables that are not pre-wrapped saves the resources that made the wrapping. Bringing your own grocery bags to a store instead of getting new bags with each visit saves paper and plastic.

Reusing items conserves resources and decreases the amount of trash. Paper plates and cups are thrown in the trash after one use, but plastic or porcelain dishes and glass cups can be used and washed again and again. You can reuse a water bottle many times to conserve the energy resources and the plastic needed to make a new plastic bottle.

Glass jars, plastic bottles, aluminum cans, and newspaper can be recycled. Aluminum cans are melted down and can be made into new objects. Plastic bottles can be melted down to make new plastic bottles. When you recycle materials that came from the geosphere or biosphere, you help prevent these systems from being mined for more resources. A compost pile made with your yard waste and kitchen scraps can be recycled into fertilizer to make the soil richer. This reduces the amount of trash you are putting in a landfill.

You can help conserve material and energy resources to protect Earth systems by reducing, reusing, and recycling. These aluminum cans have been crushed and baled at a recycling center. The cans will be shipped to an aluminum foundry to be made into new cans.

Think of some ways you can protect the environment. Write a pledge to the environment. Be specific! Include these terms: **conserve, protect, reduce, reuse,** and **recycle**.

I will . .

I will . .

I will . .

3. Engineers Can Protect Earth's Systems

Many years ago, scientists suspected that power plants were a major source of air pollution. They tested the smoke from the smokestacks of power plants and found that it contained many harmful chemicals.

Identifying the Problem

Scientists first worked to identify the problem. They found that harmful chemicals are produced when high-sulfur coal is burned. Much of the coal that is mined in the eastern United States is high-sulfur coal. Coal that comes from the western part of the country has less sulfur. There are many minerals in coal that contain sulfur, and some of these minerals contain more sulfur and some contain less. Some minerals produce gases that cause acid rain when they are released into the atmosphere. Other forms cause smog.

Designing a Solution

Scientists found that some sulfur-containing minerals could be removed from coal. Other minerals could not. Scientists and engineers worked together to design solutions to the pollution problem. Some solutions they tested were designing better smokestacks, using only low-sulfur coal, and reducing the rate at which the coal burned. They also studied the effect of scrubbing the smoke. **Scrubbers** are devices that remove dirt and harmful pollutants from smoke. This smoke is produced by burning high-sulfur coal. The engineers tested the solutions to find places where they could improve. They compared all the solutions against each other. They also tested the solutions against previously used solutions. In this way, they decided the best way to approach the problem.

These tall smokestacks from power plants send smoke into the atmosphere. Scientists and engineers worked together to design solutions to the pollution problem such as designing better smokestacks.

Balancing the Options

Engineers had to balance the best solution with many other factors. They had to think about the locations of power plants and coal mines. They had to look at the cost of shipping low-sulfur coal to eastern plants. They had to learn about government rules for how much sulfur can be allowed in the atmosphere. They also had to think about how much electricity would be produced with each solution. Some solutions lowered the rate at which electricity was produced so much that the solution would not be effective.

Applying the Solution

Engineers decided that low-sulfur coal should be burned in power plants in the western part of the country. For eastern power plants, a different solution was needed. The engineers chose burning of high-sulfur coal combined with scrubbing as the best solution.

They set to work designing new scrubbers that remove pollution more effectively. Research in scrubber technology is still going on. Today, several different kinds of scrubbers are used. Some clean harmful chemicals and gases from smoke. Some also remove dirt and other large particles. Others even clean coal before it is burned. Scientists and engineers are working on more ways to reduce air pollution.

Engineers worked to determine the best solution among many other factors. One of their solutions involves a device called a scrubber. Scrubbers remove dirt and dust, harmful chemicals, and gases that would pollute the air if they weren't removed.

Scientists and engineers have come up with different methods to protect Earth's systems. Research one method. Then write a paragraph about what you learned. Use at least one print and one digital source. Include quotes and provide your sources.

Sources:

4. Communities Can Protect Earth's Systems

Suppose you want to start recycling newspapers and plastic bottles. If your community doesn't have a recycling program, you could try starting one in your school. You could help your community protect Earth's resources and environments.

Protecting the Geosphere

Recycling helps the geosphere because less metal has to be mined from Earth's crust and less trash builds up to pollute the land. Many communities have recycling programs. Some have recycle centers where people bring materials to be recycled. Other towns have curbside pickup of materials to be recycled. Hazardous waste centers collect batteries and paint so they do not pollute the environment.

Protecting the Biosphere

Many communities have tree-planting programs. They plant trees along sidewalks and streets. The more trees that grow in a town, the more carbon dioxide they will take up to make food. Less carbon dioxide will build up in the air. Some cities have been named Tree Cities USA because of their tree-planting programs. To be named a Tree City USA, a city must plant and take care of its trees.

Communities also pass laws protecting the environment. Some laws forbid people from digging up rare plants in parklands. Other laws regulate hunting wildlife to prevent overhunting or overpopulation. There are many zoning laws that determine where companies can build factories and office buildings.

Communities can create recycling centers to protect Earth's systems. If your community has a recycling center, you may be responsible for separating and sorting items to be recycled.

© Teachers' Curriculum Institute

Protecting the Atmosphere

Many community programs protect the atmosphere. These programs aim to conserve fossil fuels and prevent the air pollution they cause. Have you ever visited a farmer's market? Many towns and cities have outdoor markets where local farmers can sell their fruits and vegetables. By buying locally-grown food products, less gasoline is needed to transport food from farther away.

Some towns and cities build bike paths and bike lanes. They encourage people to ride bikes to work and to school on safe pathways. A new bike program that some towns are adopting is called bike sharing. Bike sharing is a transportation program that provides city-owned bikes for moving short distances within the city. Users pay a small fee to pick up a bike at a bike station and return it to any other bike station. Fewer trips in cars are needed within the city.

Protecting the Hydrosphere

Often, communities have water and wastewater treatment plants. They protect water resources and ensure that citizens have safe water. Water treatment plants take water from a water source and treat it to be sure it is safe to drink. Scientists are always researching better ways to treat water. After water has been used by people, it is returned to a wastewater treatment plant. There it is treated to remove waste products. When the water is safe, it is released back into the environment.

Local farmers sell their fruits and vegetables at farmers' markets that communities create to protect Earth's systems. Less transportation is needed when people buy local products.

Interview a family member or librarian to find out what your community does to protect the environment. Then, write a letter to your neighbor that encourages her or him to take advantage of some of these community programs. Be specific about which Earth system is being protected. Include these terms: **conserve** and **environment**.

5. Countries Can Protect Earth's Systems

One of the best ways a country can protect Earth's systems is to pass national laws. Laws can protect wildlife from extinction. For example, the U.S. government passed the Endangered Species Act in 1973. This law identifies species that might become extinct, and forbids killing or catching these species. It also provides money to create wildlife sanctuaries and refuges. A **wildlife refuge** is an area of land or water that is set aside to protect wildlife. Hunting and farming are restricted in refuges.

The Clean Air Acts of 1970 and 1990 set dates for reducing the amounts of certain air pollutants. Vehicles and factories are tested to make sure they do not pass their limits. Companies are fined if they do.

Countries can also work together to protect Earth's systems. In 1997, the United Nations called a meeting of countries to find ways to prevent climate change on Earth. The result of the meeting was the Kyoto Protocol. This agreement became effective in 2005. It sets limits to the amounts of greenhouse gases each country can produce. Remember that carbon dioxide and methane are two greenhouse gases. They absorb radiation emitted by the surface of the Earth and make the Earth warmer. Climate change occurs if the atmosphere warms too much. Each country agreed to reduce air pollution by a certain amount. The United States has not agreed to these limits.

Countries can pass laws to protect Earth's systems. One law provides money to create wildlife refuges. These bison in Yellowstone National Park are protected because they cannot be hunted or captured.

Research what countries are doing to protect Earth systems. You may want to start your research with the U.S. Environmental Protection Agency or the United Nations Environment Programme. Write a report on a law or regulation that helps protect Earth systems. Include quotes and provide your sources.

Sources:

Show What You Know

Jasmine lives in an area where cars and factories produce a lot of air pollution. She wants to find information on how she can solve this problem.

Write a letter to Jasmine. Your letter should include:
- a salutation, such as *Dear Jasmine,*.
- one example of how a community or person has tried to solve the problem. Make sure to include the source of this example.
- an explanation of where Jasmine can find more information on how other communities or individuals have tried to solve the problem.
- a closing line, such as *Your friend,*.

Making Sense of the Phenomenon

Let's revisit the phenomenon: *Because of pollution, you can't drink this water.*

Think about:
- Which Earth system is being harmed?
- How can we protect this Earth system?

Use your findings from the investigation to answer this question: *How can we minimize pollution in order to protect our Earth systems?*

Claim	
Evidence	
Reasoning	

☑ Go back to page 4 and fill out the unit checkpoint for this lesson.

Performance Assessment:
Creating a Public Service Announcement About Water in Your Community

Research how your local community protects the resources and environment near where you live. Then, create a public service announcement to explain what people can do to help your community.

You will:
- obtain information about how your local community protects its resources or environment.
- examine evidence to help you brainstorm ways you can help protect the environment.
- develop a public service announcement to help inform people about how they can help protect your community.

Researching How Your Community Deals with Water Issues

While Tulare County's water issues were extreme, other communities must solve water issues, too. Each community must decide where their water will come from and how they can store this water for future use.

Research information about your community's water. Be sure to research answers to the following questions. With each answer, include the source(s) where you found it.

Where does your community get its water from?

How does your community store water for the future? Does this type of storage effect the environment around it?

What threats do your water sources face? Are these threats caused by people?

Finding a Solution to Water Issues

Model one way that people in your community can protect these water sources. Then, answer the following questions:

What materials will people need for this solution? How will these components interact to solve the problem?

How does this solution interact with the environment? Will it negatively impact any of Earth's spheres?

Creating a Public Service Announcement

Let's make a PSA that explains how your community deals with its water problems. Decide as a group what kind of PSA you will make. Create an interesting and engaging rough draft of your PSA.

Your PSA should include a description of your area's water sources, the problem it is facing, and a model of the solution that you came up with.

Your PSA should clearly answer the research questions from the **Researching How Your Community Deals with Water Issues** section.

A

air mass   A large quantity of air that has similar temperature, moisture, and pressure all through it.

air pressure   How much the air pushes on any surface.

atmosphere   The Earth system that is made up of a mixture of gases that is air.

B

biosphere   The Earth system that includes all the living things found on Earth.

C

climate   The general weather of a place over a long period of time, such as many years.

conservation   The wise use of material and energy resources. Many conservation efforts involve reducing the amount of material and energy resources used.

D

decompose   When a material is broken down into smaller pieces.

deposition   The dropping of weathered material in one place.

E

erosion   The loosening and carrying away of weathered material from one place to another.

G

geosphere   The Earth system that is made up of a thin surface layer of rock, soil, and sediments as well as the materials that are inside Earth.

H

hydrosphere   The Earth system that includes all of Earth's water.

L

landform   A natural structure on Earth's surface.

P

precipitation   Water that falls to Earth's surface. Precipitation falls in many forms, including rain and snow.

pollution   The presence of anything in the environment that can harm living things or damage natural resources.

prevailing wind   Wind that usually blows more often from one direction than from any other direction.

R

recycling   Taking materials from old discarded objects and making new objects from them.

S

scrubbers   Devices that remove dirt and harmful pollutants from smoke produced by burning high-sulfur coal.

sediments   Tiny pieces of sand, rock, and shells that settle to the bottom in layers.

T

toxic   Capable of causing injury or death to an organism.

W

water cycle   The continual movement of water between the air and the land.

water vapor   Water when it is in the gas state.

weather   The condition of the atmosphere at a place for a short period of time, such as a few hours or days.

weathering   The breaking down of rock by interactions with Earth's systems.

wildlife refuge   An area of land or of land and water that is set aside for the protection of wildlife.

Cover:
Ethan Daniels/Shutterstock

Title Page:
NASA/JPL-Caltech/Univ.of Ariz.

Unit Opener
2-3: Shutterstock

Lesson 1

6: Thinkstock 7: pond5
10: Thinkstock 11: Thinkstock
13: Thinkstock 15T: Thinkstock
15B: Thinkstock 16: Thinkstock
18T: Thinkstock 19: Thinkstock
21: Thinkstock 24: Thinkstock
25: Thinkstock 28T: Thinkstock
28C: Thinkstock 28B: Thinkstock

Lesson 2

32: Thinkstock 33: pond5 37: Patryk Kosmider 39: Thinkstock
40: Thinkstock 42T: Thinkstock
42B: Thinkstock 43: Roger Wissmann 45: Thinkstock 46: Thinkstock 48: iStockphoto 49T: Thinkstock

Lesson 3

52: Thinkstock 53: Pond5
54TL: Elvele Images Ltd /
Alamy 54TLC: Thinkstock
54TRC: Thinkstock 54TR: Craig
Doros/Dreamstime 54BL: 13claudio13/Dreamstime 54BLC: wilar
54BRC: National Geographic
Image Collection / Alamy
54BR: Ockra 59: Thinkstock
60TL: Thinkstock 60TR: Thinkstock 60BL: Westend61 GmbH /
Alamy 60BR: Thinkstock
61: Thinkstock 62: Thinkstock
64T: Thinkstock 65: Thinkstock
64B: Westend61 GmbH/Alamy
67: Thinkstock

Performance Assessment

72: Shutterstock

Lesson 4

78: Thinkstock 79: pond5
85: Thinkstock 87: koko-tewan
88T: Thinkstock 88B: Thinkstock
90: Thinkstock 91: Thinkstock
93: Thinkstock 95: Thinkstock

Lesson 5

98: Thinkstock 99: Pond5
102: Thinkstock 103L: Thinkstock
103C: Thinkstock 103R: Thinkstock 104: Thinkstock 105: Paul
Doyle/Alamy 106: Thinkstock
107: Thinkstock

Lesson 6

112: iStockphoto 113: Pond5
120: Thinkstock 121: Thinkstock
121: Thinkstock 121: Thinkstock
121: Thinkstock 121: Thinkstock 122T: Thinkstock
122B: Thinkstock 123: Thinkstock
125: Thinkstock 126: Thinkstock
128: Thinkstock 129: Thinkstock
131: Thinkstock

Performance Assessment

136: Shutterstock